Geri Schneider Winters

Kindle Best-selling author "Why Agile is Failing at Large Companies (and what you can do so it won't fail at yours)"

21 Productivity Hacks for Knowledge Workers

Ty yn Goch
Forrest
Publications
Albion, California

Published by Ty yn Goch Forrest Publications in 2016
First edition; First printing

Design and writing ©2016 Geri Schneider Winters
Cover design by Geri Schneider Winters
Author cover photo by Landers Photography, San Antonio, Texas
Quote from blog OvercomingBias.com in section 13 is copyright Robin Hanson. Used with permission of Robin Hanson.

All rights reserved. No part of this book may be reproduced or transmitted in any form or by any means, including but not limited to information storage and retrieval systems, electronic, mechanical, photocopy, recording, etc. without written permission from the copyright holder.

The only exception is that reviewers may quote brief passages for review purposes only.

Any suggestions in this book are based on the author's own experience. They may or may not be relevant to your particular situation. The author and publisher do not guarantee any results you may achieve by following the suggestions in this book. Examples and stories are typically composites of several similar events and do not describe a specific person or company.

ISBN: 978-0-9967426-5-8

For additional book resources or to contact the author go to
www.geriwinters.com

As Always

For Jason

This book is also for you who are looking
for ways to get more done in the day

Contents

Acknowledgements 9

Introduction 11
Finding More Time for What You Want to Do 12

Reducing Busy Work and Wasted Time 15
1. Plan Your Work 15
2. Urgent Things are often not Urgent 16
3. Automate Busy Work 18
4. Delegate 18
5. Avoid Meetings 20
6. Limit the Time you spend Messaging 21
7. Dealing with Phone Calls 21
8. Limit Your Online Time 23
9. Find Your Personal Time Wasters 24

Doing Productive Work more Efficiently..........................27
10. Work on One Thing at a Time 27
11. Avoid Frequent Interruptions 28
12. Limit the Time to do some Particular Task.......................... 30
13. Take Breaks 31
14. Find Your Best Working Style 32
15. Get Enough Sleep for You 34
16. Create a System, Process, or Checklist for Routine Work 35
17. Avoid Automation that Slows You Down 36
18. Do Not Schedule Every Hour of the Day 37
19. Manage Your Online Availability throughout the Day, Night, and Weekends 38
20. Work to Your Strengths... 40

21. Avoid working more than 40 hours a week 41

Bonus Hack 43

Take Steps to Reduce Anxiety and Worry 43

Acknowledgements

Super big thanks to Leigh St. John, friend and coach extraordinaire! All alone it is hard to stick with a big project such as writing a book. Leigh kept me accountable for my deadlines in a most supportive, non-judgmental way. This book could have remained on my hard drive (like the 18 other books I just have not quite finished writing). You have this book in your hands because of Leigh.

Introduction

This book is for you if you do knowledge work, that is, you think for a living. You deal with information and do work such as analysis, design, programming, data modeling, and research. You probably sit behind a computer a lot, whether in the office, in a hotel, or at the beach. People who could use the information in this book include people who:

- architect buildings, cities, software
- design bridges, tunnels, robots
- do analysis of all kinds
- program computers
- create or customize project methodologies
- do scientific research
- and many, many other kinds of work

If most of your work occurs inside your head, this book is for you.

I assume you are reading this book because you want to remove busy work and do your productive work more efficiently. But it is possible that you are in a situation where being more productive is not to your benefit. In that case, you could do the opposite of the suggestions in this book to be sure you always look busy.

Finding More Time for What You Want to Do

Busy just means you are doing things. Anything, any activity. Productive means you are doing work to produce something. Productive activity is related to what you are supposed to be producing. Productivity refers to how much you produce in some time period that is of high enough quality to be delivered. Manufacturing tends to focus on units per hour, but for brain work we might talk about how much was completed on average per day or week or even longer time periods.

When you work primarily with your brain, there are two different kinds of productivity hacks. One kind is identifying busy work and reducing it so you have more time for productive work. The other kind of productivity hack is how to be more efficient when doing productive work.

There are many things you may do that are taking your time away from productive work and many things that make your productive work less efficient. You may be anywhere from 10-75% less productive than you could be. What this means is that producing something takes far longer than it needs to.

Would you like to have 10% more time? How about 20%? 50%? Imagine what you could do if your work could be done in half the time it is now! Perhaps you can complete your current project early. Maybe you complete two designs in the time you used to take for one. Instead of producing more, perhaps you use the time to take a vacation or spend more time with your children. What you do with the extra time is up to you. What I want to do is show you how to do the same work or more in less time so you can use your new free time for the things that are important to you.

The first section of this book discusses hacks for getting rid of busy work and time wasters, giving you time to do what you want to do. The second section contains hacks for using your productive time more efficiently so you get your productive work done faster.

That allows you to produce more in the same time or produce the same amount in less time so you have more time for other things you want to do. Finally, look for the "bonus hack" at the very end.

Section 1

Reducing Busy Work and Wasted Time

1. Plan Your Work

This is so important! It is the best tool you have for removing wasted time from your day. Some people like to plan their day the night before. Others take time at the beginning of the day to decide what they will complete that day. Of course you do medium and long range planning as well, but you need daily plans to keep your focus on what is important.

When I say plan, I do not mean you have to fill up your calendar with something every hour. I am talking about something much simpler. Identify one thing each day that is the most important thing you need to complete that day and then get rid of anything that would get in the way of achieving that goal.

Consider the various things you are working on. What is the most important one you need to finish today? What might keep you from completing it? What work that you might do today can you work on some other day in order to get this most important thing done? What can you delegate to someone else in order to get this most important thing done? Do you really need to personally attend all the meetings on your calendar today? Would reviewing notes or recordings be sufficient? Can you send someone else who can report back to you?

Many times when I have something really important to complete, I have gone for 2-3 days without reading email. I did not plan to avoid email; I was just too busy on something important to think about it. The consequence of ignoring email for several days? None. Most of us cannot ignore email forever because we have other people we need to work with. But 1-2 days is not usually a problem. If someone has to urgently reach you, they will probably call or text.

There are many meetings I am invited to where I am not a participant. If I am only attending to listen, I will often review notes or a recording later instead of attending live. Since I can skim the notes or fast forward a recording, this takes a lot less of my time. The time I could have been in a meeting I spend completing my most important task of the day.

> *Productivity Hack: Pick one important thing to focus on each day. You will easily and almost automatically avoid busy work during your day if you keep your focus on that one important thing you need to complete today.*

2. Urgent Things are often not Urgent

Do you plan your work based only on what is most urgent or with the earliest due date? Urgent (I need it today!) is not the same thing as important. Urgent work may not be productive work. Deadlines are often artificial.

Before filling your time with urgent tasks, first consider how important this urgent work is. Productive work is more important than busy work. If the urgent task is to help finish planning Susan's retirement party, that is much less important than completing a report due to your boss.

When considering how important the urgent work is, think about the consequences of missing a deadline. If I do not pay taxes on time, there are penalties. If I do not complete a template today, then I hold up the work of 20 other people who are depending on it. If I do not complete my status report today, my boss will ask me for it during my one-on-one tomorrow. This is clearly much less important than the other two.

If you have a business partner or co-worker who often asks you for help because they are late on meeting a deadline, you have to question the importance of helping them versus other work you need to do. If you keep helping them, you are now doing their work instead of your important work.

Consider if you personally need to do this urgent work. Someone may have asked you to do it because you were available, but that does not mean you are the right person to do it. You may be doing urgent work out of habit without considering there may be better uses of your time.

Business Analysts are often asked to help with creating urgently needed reports because they are usually good at putting information together in a meaningful way. But that can mean the Business Analyst is not getting their important work done. If you are in the situation where someone has asked you to help but you have more important things to do, find someone you can delegate the time sensitive work to.

Finally, ask yourself if the urgent work is really urgent. Is the deadline real, or some date someone picked randomly? If there are no real consequences to delaying the work, then it is not truly urgent.

> *Productivity hack: Organize your work as Stephen R. Covey suggests in "The 7 Habits of Highly Effective People": 1 – Important and Urgent, 2 – Important and not Urgent, 3 – not Important and Urgent, 4 – not Important and not Urgent.*

3. Automate Busy Work

Many repetitive tasks can be automated to greatly reduce the time you spend on them. Since many of these tasks are busy work, this frees up your time to do productive work.

Are you spending a lot of time creating a weekly status report? Find a way to generate it automatically, perhaps using macros in a spreadsheet tool. Then you can just push a button to create the report. Or perhaps you can track your work in an online tool and send a screen shot for status.

Do you spend a lot of time moving email into categories? Put filters on your email inbox to sort incoming mail into categories automatically.

Do you watch the clock when an important event is coming up? Create alarms for important events so you can focus on your important work without watching the clock.

> *Productivity hack: Pay attention to how you spend your time and look for ways to automate repetitive tasks.*

4. Delegate

A powerful way to free up more of your own time is to delegate work to someone else. This is the basis of Tim Ferriss' famous "The 4-Hour Work Week". Identify what you personally have to do and delegate the rest of your work to someone else. Do this at work and at home.

The more senior position you want in your company, the more you have to learn that you cannot do everything yourself. You get higher level, more senior positions by doing less work yourself, but providing the vision or goals for the people working for you.

What work should you delegate? Start by looking for things that you really do not want to do and find or hire someone who loves to do those things. I knew a programmer who hated writing. He convinced his boss to hire a technical writer for the team. Everyone was happy with this solution (he was not the only team member who hated writing).

Perhaps you find turning information into a presentation tedious and boring. But you know an admin who is really good at it and loves doing it. Stop creating presentations. Give the information to the admin and let him do it.

You hate housework and would rather spend the time doing things with your children. You would be happy to pay someone to clean your house so you have a couple extra hours a week with your family. Hire someone who loves cleaning to clean your house. You will both be much happier, and you have the additional time to spend with your family.

After delegating the things you do not want to do, look for things that are not important for you to do, but that need to be done anyway. Again find or hire someone to do them.

If you are a programmer, perhaps you are spending your time this week fixing bugs in your code for next week's release. You have been asked to meet with production today about work that is planned for the next 6 months. You should focus on fixing the bugs and send someone else to the meeting with production.

> *Productivity hack: Delegate things you do not want to do and things that are necessary but not important. Find someone who loves to do that work; it is a win for both of you.*

5. Avoid Meetings

Meetings are not necessarily bad, but too many meetings leaves you no time for productive work. I have actually met people who do nothing but go to one meeting after another all day every day. I do not know what they actually accomplish, but they are very busy people. If your goal is to have more time for activities meaningful to you, reduce the meetings you attend.

Working sessions are great. Getting together with others to come up with new marketing ideas, design an app, fine tune a presentation, or develop a vision are all different kinds of working sessions. You produce something during a working session. This is a valuable use of your time.

Attending training is important, but instead of working on made up exercises, try bringing your real work to the training session and apply what you are learning to important work. Now you are being productive as well as learning something new.

When you are invited to a meeting, ask yourself are you invited as a participant or an observer. If you are there merely to observe, then getting a recording or notes after the meeting is a more efficient way of receiving information than sitting through the meeting itself.

If you are thinking of having a meeting just to share information with others, think about communicating that information in a different way, a way that takes less of everyone's time. Can you (or someone working for you) update a wiki, send a newsletter, or broadcast a text message?

> *Productivity hack: As much as possible, avoid meetings where you are only an observer (not a participant). Get the information that was shared through other means.*

6. Limit the Time you spend Messaging

First, do not do email, IM, or text messaging continuously throughout the day. There are two problems with checking all the time. First email, IM, and text messages are interruptions. You stop the important work you are doing to read the email, IM, or text. Second, you will naturally respond to the email, IM, or text message. Is that really the best use of your time? Remember, this communication interrupted your important work, so answering it is also delaying your important work. These small frequent delays throughout the day can easily cause you to find your important work unfinished at the end of the day.

To avoid interruption, set aside specific times to read and respond to those messages. Consider your personal rhythms and set aside times when you will naturally want a break from your important work. Mid-morning and mid-afternoon work well for me.

Productivity hack: Set aside 2-3 times during the day to read and respond to email, IM, and text messages and ignore them the rest of the day.

7. Dealing with Phone Calls

Consider first your personal phone. You probably want your children, partner, friends, spouse, etc. to be able to reach you at work. Unless you are willing to take their calls at any time, you should turn off your personal phone during times when you do not want to be interrupted, even by them.

When your personal phone is on, you probably do not want to take calls from salespeople or folks phishing for personal information. I think all phones have caller ID, but if yours does not, really consider getting a phone that does. Then, be sure to keep your phone address book up-to-date. When a call comes, if you do not know the caller, do not answer the phone. If they really want you personally, they will leave a voice mail that you can review later. If there is no voice mail, it was not important.

To avoid being interrupted by sales or phishing calls, I have address book entries Spam1, Spam2, and so on. I have a silent "ring tone" that I associate with those addresses. At the end of the day, I look at my phone for unanswered calls and no voice mail. I look up those phone numbers on the internet, and if those numbers are reported as sales or phishing calls, I add those numbers to my Spam1, Spam2, etc. addresses. When they call again, which they do, I never hear the phone ring because it is silent. Then I do not even have to stop to look at the phone to see if I want to answer it.

Using this technique, I never deal with salespeople, people phishing for information, or people who want "just a few minutes of your time to complete this survey". The time saved may be small for each call, but it can really add up.

For your work phone, consider turning it off during periods when you do not want to be interrupted. When I say off, I really mean completely off. Not only are you more efficient, but you appear to be very busy because you are so hard to reach by phone.

You may be thinking these hacks are not something you want to do because you need to be reachable all the time. You might be a salesperson or someone in a support role. Even if you are in those kind of roles, you will want to use these hacks during times when you are "off duty".

> *Productivity hack: For your personal phone, do not answer the phone if you do not know the caller. For all phones, turn the phone off during times when you do not want to be interrupted.*

8. Limit Your Online Time

It is often important to your business, professional, or social life to spend time online reading journals, reading blogs, participating in forums, reading and writing in social networks, and attending hangouts. At the same time, it is really easy to spend hours online without even noticing how much time has gone by.

Ask yourself two questions: Do I personally need to be online or can I delegate that to others? How much time can I spend online before it interferes with more important things (work, children, partner, etc.)?

If you are online for research, ask yourself if you personally need to do that research. Is this something you can delegate to someone else, at least to do the preliminary research? It may be good enough for you to get the results of someone else's research. Or they may be able to do the initial work of finding the credible sources of information, then you spend your time on the actual research.

If you are online building your professional reputation, again ask if you personally need to do that work. There are many people and businesses who can do different kinds of online marketing for you. While you will still be involved, it will take less of your time than if you do all the work yourself.

If you just want to be online (for whatever reason), treat it like messaging. Do not have forums or social media open all day long interrupting you throughout the day. Pick 2-3 specific times when you will be online, and have those sites closed the rest of the day.

In addition, you may want to set a timer or alarm to help you limit how much time you are online. If you choose to spend 1 hour at a time online, then set an alarm for 1 hour and close the site at the end of that time. This will really help you keep from spending 2 or 3 hours online when you only meant to spend 1.

> *Productivity hack: Set aside specific times to be online. Strictly limit how much time you spend online.*

9. Find Your Personal Time Wasters

There are probably other things that waste your time throughout the day that I have not discussed. To find them, keep a time journal for a week. A time journal is just making notes of the things you do all day and how much time you spend doing them. You can try just remembering, but there will be a lot of things you forget, the common routine things you do all the time. Those are the things we particularly want to find because that is often where time is being wasted or used less efficiently.

You can do this however is easiest for you – put notes in your phone or tablet, carry a little notebook and write in it, or even dictate notes into your phone or a recorder. Since I am in front of a computer so much, I have used a cloud based calendar for this purpose. I just keep it open all day and make notes in it whenever I change tasks. However you do it, make it the easiest way you can think of, and make sure you can do it consistently. You do not need to track to the minute. I typically track my time to the nearest half hour, but no shorter than 15 minutes.

At the end of the week, you will probably find that you are more productive because you were paying attention to how you were wasting time throughout the week. Being mindful of how you spend your time causes you to use the time more productively. You can also review your time journal at the end of the week looking for wasted time and inefficient time.

You may find you are waiting a lot. What can you bring with you to the places where you wait so you can use that time for something more valuable than playing a game on your phone or reading old magazines? I find waiting periods a good time to go through email or look up an article online that I want to read. A lot of people like audio books when they are in the car for a long period. Many teachers I know take any opportunity to work on grading homework assignments. 5 minutes waiting for a teller at the bank could be 5 short assignments reviewed and graded.

> *Productivity hack: Be mindful of how you are spending your time. A time journal can assist you in finding wasted time.*

Section 2

Doing Productive Work more Efficiently

10. Work on One Thing at a Time

Focus is the most important thing you can do if you want to work the most efficiently. Many of the hacks in this section are designed to allow you to focus on your productive work. Doing one thing at a time is the single best thing you can do to be more focused.

Every time you switch between different tasks (or projects) your brain takes some amount of time to reset. Your short term memory has to unload the information it was keeping handy for task 1 then load in the information it needs to keep handy for task 2. Depending on the nature of the work you are doing, this task switching time can range from a few minutes up to 30 minutes.

You might think 30 minutes is far too much time to switch between tasks, but think about what has to happen. First you have to end your work on task 1: you make notes about what you were doing and what needs to be worked on next, send any messages required to other team members, and perhaps update information in an online tool. This can easily take 15 minutes.

Second you have to prepare to start task 2: you have to review notes about what you were doing last and what needs to be worked on next, review any messages concerning this task from other team members, and review the section of the work product you want to start to work on. This can easily take another 15 minutes.

Every time you switch from one task to another you lose some productive time. The amount of time depends on how much work you have to do to make the switch. The more often you do this during the day, the more productive time you lose.

Assume you have three projects to complete and each one will take you a full week. You will be most productive if you work on one of them until it is done. Then work on number 2 until it is done. Then finally work on number 3 until it is done. This will be 3 weeks of work total, but at the end of the first week you have completed project 1, at the end of the second week you have completed project 2, and at the end of the third week you have completed project 3.

If you work 1/3 of the day on each of the projects every day and it takes about ½ hour to switch from one to the other, you will add approximately 20 hours to the time it takes to complete the three projects, about an extra half week of work. Not only that, none of the projects will be complete until the middle of week 4.

You might be thinking that you need to switch tasks because you are stuck on one of them. You are probably stuck because you are mentally tired and need to take a short break. Try this: walk away from the work and go get something to drink. You are probably not only mentally tired but also dehydrated. A 5-10 minute break with a little walking and getting something to drink will likely bring you back to your work with fresh ideas.

> *Productivity Hack: Focus on one thing at a time until it is done.*

11. Avoid Frequent Interruptions

Closely related to multi-tasking is dealing with interruptions (a kind of multi-tasking). You are working on something and you are interrupted. It takes a moment for you to "come out" of your thoughts to put your attention on the interruption. You handle whatever the interruption requires which takes some amount of time away from your work. You return to your work and have to figure out what you were doing, what you were thinking, and what you need to do next. You may have lost ½ hour or more of working time due to that one interruption.

Have you ever spent a whole day trying to write one page because you kept getting interrupted over and over?

This is why a lot of executives have an executive assistant or chief of staff who controls access to "the boss" so that the boss can get their work done. The boss does not get interrupted. Either the executive assistant or chief of staff handles the interruption or the item that needs the boss' attention is added to a prioritized queue of work for the boss to review at specific times during the day.

You probably do not have someone guarding your time when at the office. And if you work from home, you may have pets, children, neighbors, delivery people, etc.. interrupting you. You need to find a way to hide.

Close your door. Go to a coffee shop or library. Go to another part of the building and find an out-of-the-way place to do your work. Go to a park and find a shady spot to work (so you can see your computer or tablet screen). Turn off your phone.

If you are watching your children and trying to work from home, you may find that hiring a babysitter for a couple of hours is a worthwhile investment so you can get your work done. Find a way to avoid interruptions.

> *Productivity Hack: Find a place where you can hide, physically and virtually, to get work done without interruption.*

12. Limit the Time to do some Particular Task

Especially for creative work and open-ended tasks such as analysis and research, the work tends to expand to fill the time allowed. Production book writers work to deadlines for a reason. It is all too easy to continue making the book "just a little better" and never get the book to market. Publishers know this and so set deadlines for their writers. Do the same for yourself, especially for very creative work and open-ended work. Hold yourself accountable for completing in that time period.

Challenge yourself a bit. We often estimate our work to take longer than we think it will take. When you have a boss looking over your shoulder holding you accountable for finishing "on time", it is reasonable to overestimate. But when you are accountable to just yourself, set more aggressive time limits. Otherwise you may stretch the work out to fill the time allowed, even though you could have finished sooner.

If you need to write a 350 word summary, and you allow all day, it will probably take all day. If you allow an hour, it will probably take an hour. If you are a programmer, can you code and test that component this week? If you have ever been to a hack-a-thon you know how much can be done in a short period. We tend to be more creative when given limitations than when there are no limitations. So try time as a limitation on your work.

For my most important work, I like sharing my goals with a buddy. My buddy also shares her goals with me. Just telling someone when you will have something done makes it far more likely that you will complete it by the time you said you would.

> *Productivity hack: Find a buddy and take turns holding each other accountable for getting your work done in the time you allowed.*

13. Take Breaks

When I reached this point in writing this book, I was tired and short of ideas. I was really slowing down. I looked at the clock and realized I had been working steadily for 3 hours. I was physically stiff, mentally tired, and somewhat dehydrated. So I walked away from my desk and went to get an iced tea. 10 minutes later, and I am ready to work again. It is toward the end of the work day, but I'll probably get another hour of really productive work done before I stop. Just 10 minutes has gotten the ideas flowing again.

"Not only do people need to rest and sleep at the end of a work day, but on the job mental fatigue reduces mental performance by about 0.1% per minute. Since by resting we can recover at a rate of 1% per minute, we need roughly one tenth of our workday to be break time, with the duration between breaks being not much more than an hour or two." From the post "Excess Turbulence?" by Robin Hanson at http://www.overcomingbias.com/2015/08/excess-turbulence.html accessed October 6, 2016.

I did the math for you. This works out to taking a 7 minute break per hour (not counting the last hour since you are stopping work at the end of that hour anyway). Or you could take a 15 minute break every 2 hours.

During your break, be sure to do several things to restore your energy, both physical and mental. 1. Stretch, blink, and yawn. This refreshes your eyes and gets oxygen moving to your brain. 2. Move physically away from your work for a few minutes. This tells your conscious brain to disengage and provides some mental relaxation. 3. Get something to drink. You are probably a bit dehydrated, especially if you have been focusing really intently.

To remind yourself to take breaks, you can set an alarm. If you are on a computer, there are apps that will lock your screen for a period you set, such as 5 minutes every hour, which force you to do something else. I personally do not like those, because I may be just a minute or two from a natural break point in the work, so the locked screen becomes a hugely annoying interruption. Other people really love these apps.

> *Productivity hack: Take regular short breaks throughout the day that include physical movement and getting something to drink. If you need to, set an alarm to remind you to take breaks.*

14. Find Your Best Working Style

To be the most efficient in your work, pay attention to your personal rhythms – what time of day do you do your best work? When are you the most energetic? When do you need to break to eat? Are you a mental introvert or extrovert or somewhere in the middle?

When are you at your best mentally? Do your heavy mental work at those times because you will be most efficient then. Other times of day, do things that do not require a lot of concentration or focus. I am a slow starter in the morning, so I tend to do things such as go through email first because it is usually very easy work. After that, I can sit and focus on a heavy mental task until lunch time. I get another period of good mental work about mid-afternoon. If I try to do high concentration activities first thing in the morning, such as writing this book, the work goes very slowly and I am frustrated.

You will probably notice that some times during the day you just need to get up and move. You have physical energy, but not so much mental focus. This is a good time to do more interactive work, especially if it gets you on your feet and moving around. These are good times for activities such as an idea generating session around a white board. Or perhaps taking a walk in a park, around campus, or around your office building. If you work out at the gym, your more energetic times of day are great for that purpose.

Eating also sets up a rhythm. Do you nibble all day, eat 3 regular meals, eat only breakfast and dinner, or eat only once? If you change the rhythm of meals, what does that do to your energy, both mental and physical? You may find that if you do not eat regularly, you get shaky and have trouble concentrating. That will make you less efficient at your work. Are you getting enough liquids to drink? Dehydration leads to mental slow down, so if you are having a hard time concentrating, try getting something to drink.

Consider also how your brain can work at its best. Science explains introvert and extrovert a little differently than you may have heard. Our brains have an optimal level of activity when we are most efficient. Activity in an extrovert brain is below that optimal level. Activity in an introvert brain is above that optimal level.

An extrovert needs outside stimulation to bring their brain activity up to the point where their brain works the best. If you have an extrovert brain, you work best when there are things going on around you or you can listen to music. Of all people, you are most likely to multi-task because changing what you are working on frequently helps bring your brain activity up. Working alone in a quiet room just slows you down more, leaving you feeling dull and uncreative.

An introvert needs lack of stimulation to bring their brain activity down to the point where their brain works the best. If you have an introvert brain, you work best working alone in a quiet space. Of all people, you are least likely to multi-task because it is too distracting. Working in an area full of activity and sound just speeds you up more, leaving you feeling unfocused and nervous.

> *Productivity hack: Figure out your best working style and arrange you work environment and schedule to match.*

15. Get Enough Sleep for You

If you are not getting the right amount of sleep for you, then you will be less efficient at your work. You know this, but you try to get by on less sleep to get more done. Instead of cutting back on your sleep, try applying the productivity hacks in this book to recover some of your wasted time. You really do need to get enough sleep.

How many hours do you need to sleep? Science provides a guideline of 8 hours a night, but that is an average. Some people really do need less and some really do need more. You probably have a pretty good idea of the right amount of sleep for yourself; the problem is more that you do not always get it.

Many people seem to have a subconscious feeling that sleep is just wasted time. You need to convince yourself that if you get the right amount of sleep, you will work faster and better. Therefore, sleep is not wasted time at all.

When considering sleep, think about not just how many hours, but when you get your best sleep. Some people sleep once in a 24-hour period (monophasic). Some sleep twice per day (biphasic). And others sleep many times throughout the day (polyphasic). Biphasic is a quite common pattern for humans, where you do most of your sleeping at night, but have a nap (siesta) in the afternoon (when most of us are drowsy anyway).

If you know your sleep pattern, then you can figure out the best time(s) of day for you to sleep. The majority of people do most of their sleeping at night. But this is certainly not 100% true. Knowledge workers generally have more flexibility in their schedule than many other workers, so you can take advantage of that to sleep the hours that are most natural for you.

Hours of sleep is one part of the equation, but quality of sleep is another. There is a lot of research on the best environment for good sleep, but again, you know yourself best. Some people like to be cold, some people like to be hot. Some people require absolute quiet, some need a white noise machine, and others need city sounds. Everyone sleeps best in a familiar environment. One reason for frequent travelers to use the same hotel chain for every trip is that the hotels are similar enough that they are a familiar environment for the traveler.

> *Productivity hack: Determin how much sleep you need, what time you need to sleep, and what your environment needs to be. Follow that pattern to get the sleep you need.*

16. Create a System, Process, or Checklist for Routine Work

What do you do over and over the same every time? If you create a system or process for that set of activities, and do them the same way every time, then you will become very efficient at doing them. As your system or process becomes habit, you do not have to worry about forgetting something.

This is an especially good thing to do for tasks that prepare you to be productive and the tasks you do to clean up after your productive work is done. Those are times when you are likely to be less focused and more forgetful.

Many people have a morning routine which includes all the things they do between waking up and starting productive work.

This may include things such as exercise, meditation, going through email, reading Facebook updates, making breakfast for the children, packing lunches, or reading the newspaper over a cup of coffee.

If you plan to write a report today, what are all the things you need to do? Research, analysis, information design, and final review might be the usual tasks. If you write reports often, you could keep a checklist of all the things you need to do. You might think you do not need to because the work is obvious. One Business Analyst I worked with said he always forgot the final step, which was to make a copy of the report to a specific server where archives were kept. It is those little things you always forget that are perfect for a checklist.

When I stage manage musicals, I have a routine at the end of every night before I go home. I walk around the inside of the theater building making sure lights are turned off, doors are locked, everything is put in its place, and the trash is picked up. Because it is a completely mechanical routine (I walk the building and look), I never forget even when I am very tired.

> *Productivity hack: Turn a routine sets of tasks into a system, process, or checklist that over time becomes habit.*

17. Avoid Automation that Slows You Down

Automation is wonderful, software is wonderful, but sometimes it causes more problems than it solves. Sometimes when you use that cool software app, you spend more time doing the job than if you did it a more manual way.

Perhaps you are designing something. You could design it in an app, which takes a long time. Or you could draw your designs on a whiteboard, then when you are happy, you could photograph the designs and share the photos with whoever needs to see the designs. Use the app when it becomes necessary, for example when you have completed the final design and now need to model your idea in CAD to send it to a 3-D printer.

You are in an idea generating meeting and someone connects their computer to a projector to share a document. As everyone shouts out ideas, the person behind the computer types the ideas into the document. This is incredibly slow compared to giving everyone a pad of sticky notes and asking them to write one idea per sticky note and put them on the wall. Or if there are a lot of white boards, then have everyone write ideas on the white boards. You can photograph the results and post online. If it becomes necessary to manage the ideas (perhaps they have become requirements for a project), then they can be data entered into whatever tool is required.

One of the greatest inventions is Idea Paint. This paint turns any wall into a white board. Imagine if all the walls in your office were white boards and you could use any dry erase marker to jot ideas on the wall as they occur to you. Then you could easily erase and change those you do not want. Doing everything on a computer is not always the most efficient approach.

Productivity hack: If automation is slowing you down, replace it with something more efficient.

18. Do Not Schedule Every Hour of the Day

Do not schedule every hour of the day for productive work. Block out at least 2 hours a day in your calendar with nothing specific planned for that time. You can do this by making appointments with yourself so others will not fill this time. The reason for doing this is that unexpected things always occur. By not planning 2 hours a day, you have time to deal with those unexpected things and still get the work done that you planned.

You will be doing things during those 2 hours a day! Things such as figuring out the next most important thing to work on. Or an impromptu design conversation over coffee. Or answering questions from a new team member. Or finding out that what you thought would take 2 hours took 4 hours. Or creating the report everyone forgot has to go to the Board of Directors tomorrow.

If you try to schedule 40 hours of productive work a week, then you will consistently work more than 40 hours a week doing all those other things that came up in addition to the 40 hours of work you scheduled. Over time, you will become more and more exhausted.

> *Productivity hack: Keep 2 hours a day unplanned so you have time to deal with the unexpected throughout the day.*

19. Manage Your Online Availability throughout the Day, Night, and Weekends

With home computers and smart phones, many of us are finding there is no time away from work. If you manage it right, this can let you put in fewer hours, be more productive during those hours, but look like you are working long hours. This can be useful if you work for a company that expects everyone to work long hours. If you do not manage your online availability, you never have time away from work.

For most kinds of knowledge work, you do not really have to instantly answer a phone call, text, or email from work. Corporate trainers routinely ask students to turn off their phones and only check messages during breaks. Important meetings are another time when you do not want to answer the phone. You do not want to be interrupted when deep in a creative endeavor. Maybe you do not want to be interrupted at your daughter's basketball game, or maybe you do not mind taking calls then.

The basic strategy is to decide when you are OK with work interrupting your life and when you are not. The traditional approach was to go into the office to work from 8am - 5pm and only work when at the office. For a really large number of people, that is not possible anymore. So pick times of day or specific events when you will not answer messages or phone calls from work. At the end of that time period, you can check for messages and answer them.

You might say that 6pm – 9 pm is family time when you do not allow work interruptions. Then you check messages at 9pm and spend no more than ½ hour answering them. Perhaps you check messages at 6am, and by 6:30am you are at the gym working out. You are not available for work until 8am. 9am-11:30 you might be focused on productive work and allow no interruptions. Then you check messages before lunch.

If you wake up during the night with an idea to email the boss, you can do that. You could answer other messages at that time as well. Or you might say that 8 hours a night is yours, and if you are awake, you will not spend the time thinking of work.

By identifying time periodds when you will not work, you can mingle work and personal time throughout the day and weekend if that is what you want to do. You just have to watch how much time you are actually spending working so you do have time away from work to do other things.

> *Productivity hack: Pick the times and events when you will be available to do work and when you will not.*

20. Work to Your Strengths

You will be most efficient when you do things you are skilled at. So to be the most productive, choose tasks that are in your areas of expertise.

This is not to say you should never learn something new, but it does mean that you will be less productive when you are learning new things. In that situation, allow yourself more time to get the job done than you would otherwise. It seems obvious I know, but many people just do not think about the extra time they need to learn new things.

Think also about the manner in which you work. You will be most productive when you work in a manner that is comfortable for you. If you need routine and structure, you will not be comfortable in an environment where everyone is encouraged to challenge the status quo. If your nature is to be extremely honest and clear when communicating, then you will not be comfortable in an environment where successful people hide information. If you really like working collaboratively, you will not like an environment where everyone works alone in a cubicle. You cannot be at your best when you have to work in a manner that is not natural to you.

> *Productivity hack: Pick work to do that you are good at and that allows you to work in a manner where you are most comfortable.*

21. Avoid working more than 40 hours a week

Productivity (the amount of output produced per day) generally goes down as hours worked increases. If you work much over 40 hours a week for an extended period of time, your total output will be less than if you worked a steady 40 hours a week. You get negative productivity when working much over 40 hours a week for long periods.

There are no studies showing that working more than 50 hours per week for months on end leads to higher total output. There are a lot of studies showing that if you cut the work week from 50 hours to 40 hours, productivity increases. You will get more done in 40 hours a week than you will in 50 (or more).

Knowledge work requires intense concentration and mental exertion. As you get more and more mentally tired, you make more mistakes and the mistakes are more costly. In addition, because you are mentally tired, the mistakes tend to take longer to fix. Then you work longer hours to catch up your work, making you even more tired, and leading to more mistakes.

You are going to tell me you know someone who always works way more than 40 hours a week and is super productive. There are many ways to give the appearance that you are working far more hours than you actually are. If you cannot think of any, an internet search will provide numerous ideas. Your overachiever is probably doing a number of things that make it appear they are working a lot more hours than they actually are.

You can work more than 40 hours a week for about 3 weeks and get a short-term boost in productivity. After 3 weeks, your productivity will quickly drop and you will be far less productive than before.

If you do work long weeks, do it for no longer than 3 weeks. At the end of the three weeks, cut your work week down to 20 hours for the next week to let your brain recover. This is not vacation time, it is comp (compensation) time for the extra hours you worked over the previous 3 weeks.

> *Productivity Hack: Be "on-the-job" no more than 40 hours a week. That includes onsite and from home.*

Section 3

Bonus Hack

Take Steps to Reduce Anxiety and Worry

When you are anxious or worried, it affects your productivity in two ways. First, you will find it hard to get started on anything and therefore will tend to fill your time with busy work. Second, when you are working on something productive, you will tend to make more mistakes and have trouble concentrating on the job. These things happen because when you are anxious your body and brain are in fight or flight mode. Your adrenaline is raised and you are somewhat tense.

To do your best work, you want to be relaxed and focused, in the flow, which is the opposite of how you feel when anxious or worried.

Many people tell me, "I am not worried about anything, I am not anxious". But they act like they are. This tells me that they have some constant low-level anxiety that they are not recognizing. You can do an online search for anxiety and find all kinds of information about how anxiety is extremely common and, for a lot of people, mostly continuous. A lot of people have a lot of subconscious worry all the time.

So what can you do about it? In the short term, learn the signs of anxious behavior. When you notice yourself acting as if you were anxious, do not worry about whether you are worrying or not, just take steps to reduce or remove the anxious behavior. Some things that may indicate anxiety are: anger over minor issues or problems, tense muscles especially in your shoulders and back, inability to focus, mind keeps running from one thought to another, fidgeting or restlessness, quick and shallow breathing, elevated heart rate (at rest), and palpitations.

Here are some ideas you can try if you are feeling anxious right now:

- Have a cup of chamomile tea
- Try a protein snack (nuts, an egg, some sliced turkey)
- If you are feeling chilled, do something to get warm
- Breathe lavender scent (I keep lavender lip balm for chapped lips and to smell if I am feeling anxious)
- Yawn
- Watch your breathing; be sure your abdomen moves in and out as you breathe (not your chest)
- Try 4-7-8 yoga breathing – inhale through your nose for a count of 4, hold your breath for a count of 7, then exhale through your mouth for a count of 8
- Go for a short 10-20 minute walk (with no phone or tablet) among plants and trees, paying attention to the sights, sounds, and smells around you
- Stretch
- Listen to a short guided meditation
- Practice mindful awareness of the present moment

> *Productivity Hack: Take steps to reduce the symptoms of anxiety when you want to be most productive. If you can, identify what is making you anxious and find mitigations that reduce that anxiety.*

About the Author

Geri Schneider Winters is a polymath with a wide range of interests. She loves bringing all that knowledge to bear when solving large, complex problems. Because of that, she is frequently found guiding business transformations at large companies.

In support of her business transformation work, Ms. Winters has studied and put into practice domains such as analysis, science of the brain, hypnosis, psychology, influence and advocacy, anthropology, philosophy, adult education, communication, marketing, interviewing, and a wide range of documentation techniques.

Ms. Winters explores her creative side with hobbies in healthy living, home brewing, cooking, photography, writing, book publishing, website creation, video production, singing, acting, and musical theater production. She has a deep love of the natural sciences and has been known to read physics "for fun", but admits to being "horrible" at tennis, basketball, and statistics.

Ms. Winters lives in a redwood forest on the Northern California coast with her husband and cats. She shares her property with deer, bunnies, skunks, foxes, many kinds of birds, and at least one bobcat. She is also within the territory of a mountain lion.

Other Books by this Author

The original edition of Applying Use Cases was the first published book devoted to the topic of use cases - and an instant best seller. Schneider and Winters showed us not only how to write use cases, but what to do with them throughout a full incremental development lifecycle. The second edition was updated to UML 2.0 and expanded to show how to write use cases for business, and how to flow business use cases into software.

Applying Use Cases: A Practical Guide has been used in professional training in business analysis, Agile development, software architecture, and project management. It has also been the required text for project management courses at many universities.

This popular book has been continuously in print worldwide for over 15 years. It is available from Addision-Wesley Professional in US English, Polish, and Japanese editions.

Other Books by this Author

Geri Schneider Winters
Best-selling author "Applying Use Cases: A Practical Guide"

Why Agile is *Failing* at Large Companies

(and what you can do so it won't fail at yours)

This popular book on Agile debunks the many stories about how easy it is to bring Agile software development into large established companies. The truth is that changing your software development practices to be one flavor or another of Agile may be a bad thing for other parts of your company. Changing those other parts of your company so it all fits together again may be so expensive that the return on the investment is not worth it.

Before jumping on the Agile bandwagon, before starting down the path of tearing your company apart and rebuilding it, spend a little time investigating how big the change might be and if it will be worth it.

This 2015 book has been an Amazon Kindle best seller in Organizational Behavior, Problem Solving, and Management and Leadership.

Other Books in the Productivity Hacks Series

Geri Schneider Winters
Kindle Best-selling author "Why Agile is Failing at Large Companies (and what you can do so it won't fail at yours)"

21 Productivity Hacks for Company Founders

Geri Schneider Winters
Kindle Best-selling author "Why Agile is Failing at Large Companies (and what you can do so it won't fail at yours)"

21 Productivity Hacks for Freelance Writers

www.ingramcontent.com/pod-product-compliance
Lightning Source LLC
Chambersburg PA
CBHW070553300426
44113CB00011B/1891